A. Stanford Morton

Refraction of the Eye

its diagnosis and the correction of its errors, with a chapter on

keratoscopy

A. Stanford Morton

Refraction of the Eye

its diagnosis and the correction of its errors, with a chapter on keratoscopy

ISBN/EAN: 9783337406370

Printed in Europe, USA, Canada, Australia, Japan

Cover: Foto ©Andreas Hilbeck / pixelio.de

More available books at **www.hansebooks.com**

REFRACTION OF THE EYE

ITS DIAGNOSIS

AND THE

CORRECTION OF ITS ERRORS

WITH CHAPTER ON

KERATOSCOPY

BY

A. STANFORD MORTON, M.B., F.R.C.S. Ed.

SENIOR ASSISTANT SURGEON, ROYAL SOUTH LONDON OPHTHALMIC HOSPITAL; CLINICAL
ASSISTANT, MOORFIELDS OPHTHALMIC HOSPITAL.

LONDON
H. K. LEWIS, 136 GOWER STREET.
1881

LONDON
PRINTED BY H. K. LEWIS,
136 GOWER-ST., W.C.

PREFACE.

THESE notes are published with a view of enabling Practitioners to diagnose, and correctly estimate the value of, the phenomena indicating the state of a patient's refraction. They are intended to furnish a basis for observation: and it is hoped that they will make evident the necessity which exists for *personally working out* a large number of refraction cases in order to acquire anything like proficiency in prescribing correct glasses.

To those friends who have aided me by their suggestions I would take this opportunity of expressing my best thanks.

Welbeck Street, W.
March, 1881.

TABLE OF CONTENTS.

REFRACTION OF THE EYE.

INTRODUCTION.

THESE pages are intended for beginners, and for those, such as Physicians or general Practitioners, who, systematically using the Ophthalmoscope in their investigation of disease, wish to avail themselves of the information thereby afforded regarding the patient's refraction, the errors of which they must be able to detect in order to make due allowance for them. If the patient can assist by his answers, many valuable indications are afforded by the use of test-types and glasses, a description of which, with the information derived therefrom, is therefore introduced.

Some hints are given as to glasses required in the more *ordinary* cases, but for the various diseases which accompany and complicate many of the errors of refraction and modify the ordering of spectacles, the reader is referred to the works already published on this subject, to the study of which these notes are intended to prepare the way.

So essential is the remembrance of some of the facts hereafter mentioned, that, at the risk of tautology, they have been kept constantly before the reader. The plan which has been found so useful in other branches of medical study has been adopted here, viz. to work out from the

B

symptoms the nature of the *defect* rather than to name the defect and then describe the symptoms accompanying it.

In examining refraction it is especially necessary to proceed *systematically*. On a patient complaining of bad sight we should examine him somewhat in the following manner :—

1. Listen carefully to the nature of his complaint.

2. Test and note the *near* and *distant* vision *without* glasses.

3. Examine the refraction with the ophthalmoscope, and then, having by these means arrived at a conclusion, proceed to confirm the opinion by means of test-glasses.

Two methods of measuring refraction are at present in vogue (*vide* Chap. i.). The new system of numeration is the one which has been adopted in the text. The *approximately* corresponding measurements in *English* inches have, however, in all cases been given in *brackets* for the convenience of those who prefer the old method.

Only so much ophthalmoscopy has been introduced as was necessary to explain the phenomena of refraction.

CHAPTER I.

OLD AND NEW SYSTEMS OF MEASUREMENTS.[*]

IN the old system the lenses were numbered according to their focal lengths in inches. Their *refracting* power, being the *reverse* of their focal lengths, was represented by a *fraction*, of which the numerator was 1 and the denominator was the focal length in inches. Thus, a lens of 6 ins. focus had a refracting power of $\frac{1}{6}$, *i.e.* $\frac{1}{6}$ the refracting power of a lens whose focal length was 1 in. $\left(\dfrac{1}{1 \text{ in.}}\right)$: this latter was therefore the *unit* of measuremet.

The intervals between the lenses were irregular and the difference between the refracting powers of any two lenses had to be calculated by means of *fractions*.

In the new or *Dioptric* system the *unit* of measurement is a lens whose *focal length* = 1 metre or *Dioptre* $\left(\dfrac{1 \text{ m.}}{1}\right)$ with a *refracting power* consequently $= \dfrac{1}{1 \text{ m.}}$ Two such lenses have double the refracting power or $\dfrac{2}{1 \text{ m.}} = 2$ D. with a focal length $\dfrac{1 \text{ m.}}{2}$ or $\left(\dfrac{100 \text{ cm.}}{2}\right) = 50$ cm. Ten such lenses have ten times the refracting power $\dfrac{10}{1 \text{ m.}} = 10$ D. and a focal length $\dfrac{1 \text{ m.}}{10} \left(\dfrac{100 \text{ cm.}}{10}\right) = 10$ cm.

[*] *Vide Roy. L. Ophth. Hosp. Reports*, vol. viii., p. 632, translation of an article by Dr. Landolt from which much in this chapter has been borrowed.

If we express the number of dioptres by d, and the focal length by F, then (i) To find the number of dioptres which correspond to the focal length of a lens we have the formula $d = \dfrac{1}{F}$, and (ii) For the focal length corresponding to any dioptre the formula $F = \dfrac{1}{d}$.

In order to ascertain the corresponding numbers in the two systems for any lens, we must remember that 1 dioptre (or *metre*) = 37 Paris inches and (39.4 or nearly) 40 *English inches* and that, therefore, the lens with 1 D. focal length corresponds to the lens in the old series whose focal length is 40 in. ($\frac{1}{40}$): 2 D. to a lens with double this refracting power or $\frac{2}{40} = \frac{1}{20}$ of the old.

Where d. then represents the number of dioptres and a. the number of inches, if we wish (i) to ascertain what lens in the old series corresponds to one of the new, we have the formula $\dfrac{d}{40} = \dfrac{1}{a}$. (ii) For the number of a lens in the new system corresponding to one in the old we have, $d = \dfrac{40}{a}$.

The old system of lenses in the trial cases, generally in use, have their focal lengths in inches, and do not correspond exactly with the measurements in dioptres. The following table gives the lens which, in an ordinary French or English case, will correspond *most* nearly with any dioptre.

For greater accuracy in any special case, the reader must calculate according to the formulæ given above.

TABLE.

D.	Focus in English inches.	Focus in Paris inches.	D.	Focus in English inches.	Focus in Paris inches.
0·25	160	144	5·0	8	7
0·50	80	72	5·50	$7\frac{1}{2}$	$6\frac{1}{2}$
0·75	50	50	6·0	7	6
1·0	40	36	7·0	6	5
1·25	30	30	8·0	5	$4\frac{1}{2}$
1·50	24	24	9·0	$4\frac{1}{2}$	4
1·75	22	20	10·0	4	$3\frac{3}{4}$
2·0	20	18	11·0	$3\frac{1}{2}$	$3\frac{1}{2}$
2·25	18	16	12·0	$3\frac{1}{4}$	3
2·50	16	14	13·0	3	$2\frac{3}{4}$
2·75	14	13	14·0	$2\frac{3}{4}$	$2\frac{1}{2}$
3·0	12	12	15·0	$2\frac{3}{4}$	$2\frac{1}{4}$
3·50	11	10	16·0	$2\frac{1}{2}$	$2\frac{1}{4}$
4·0	10	9	18·0	$2\frac{1}{4}$	2
4·50	9	8	20·0	2	$1\frac{3}{4}$

CHAPTER II.

ACTION OF RAYS AND LENSES.

a. RAYS of light issuing from every point of any object diverge in *all* directions.   The *nearer* an eye is to such an object the *more* of its *di*verging rays does it intercept, and *vice versa*, the further the eye is removed, the fewer are the divergent rays which reach it: *i.e.*, the more parallel are the rays which enter it.   Only from objects at an *infinite* distance do we *thoretically* get absolute parallelism of the rays.

For *practical* purposes, however, we may *consider* those *parallel* which reach the eye from an object situated at a distance of 6 metres (20 ft.) or more.

FIG. 1.

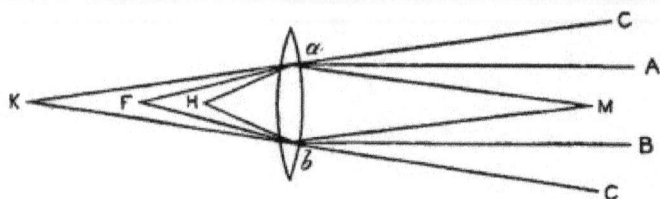

b. Parallel rays (Fig. 1., *Aa*, *Bb*), falling on a convex spherical lens come to a focus at a point, *F*, on the other side of the lens called the "principal focus."   Conversely, rays from a point at the principal focus of the lens emerge parallel.

c. Diverging rays (Fig. 1., *Ma, Mb*) meeting such a lens unite at a point, *K*, which is *behind* the principal focus. Conversely, rays from a point further from the lens than its principal focus emerge as convergent rays.

d. Converging rays (Fig. 1., *Ca, Cb*) falling on a convex lens meet at a point, *H, nearer* the lens than its principal focus. Conversely, rays from a point *within* the principal focus are, on emerging from the lens, *still* divergent, though *less* so than before meeting it.

FIG. 2.

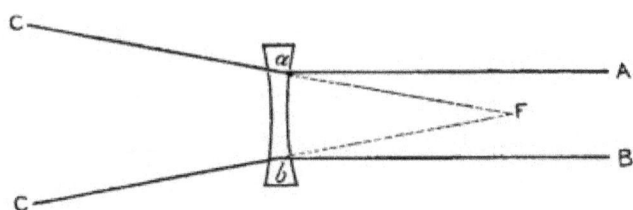

e. Parallel rays (Fig. 2., *Aa, Bb*) falling on a *concave* spherical lens acquire a *divergence* (*aC, bC*) as if proceeding from a point, *F*, on the *same* side of the lens as that from which the parallel rays proceed. The distance of this point from the lens gives the *negative* focal length of the lens. Conversely, rays with a convergence as if proceeding to this point are rendered parallel.

f. A cylindrical lens acts according to the same rules as a spherical, but *only* in a direction at *right angles* to its axis.

CHAPTER III.

DEFINITIONS.

DEF. 1. The refracting surfaces of the eye are, according to Donders—*a*, the anterior surface of the cornea; *b*, the anterior surface of the lens; *c*, the anterior surface of the vitreous. The transparent media together form the "*dioptric system*" of the eye, which has an action similar to that of a biconvex lens.

DEF. 2. "*Refraction of the eye*" means the effect which, by reason of its form and structure, this organ exerts upon rays of light entering it when *completely at rest*.

FIG. 3.

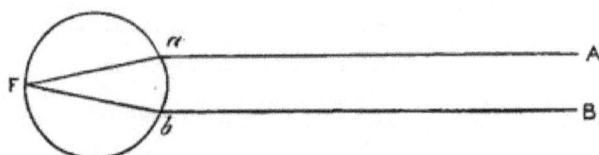

DEF. 3. In *emmetropia* (Fig. 3) *parallel* rays, *Aa*, *Bb*, come to a focus *upon* the retina, *F*, when the eye is at rest. Conversely, rays from the retina emerge parallel from the eye.

FIG. 4.

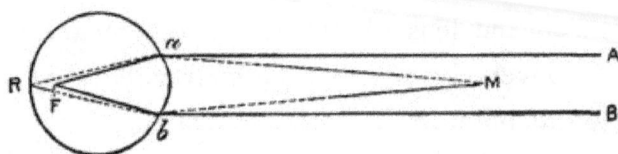

DEF. 4. A *myopic* eye (Fig. 4) is that in which, when at rest, *parallel* rays come to a focus in *front of* the retina, as at *F.* Conversely, rays issuing from the retina, *R*, which is behind the focus of the dioptric system, emerge as convergent rays (Chap. ii., *c*) which meet at a point, *M* in front of the eye. Further, it is evident that rays from this point (called the *"far point"*) would meet upon the retina.

FIG. 5.

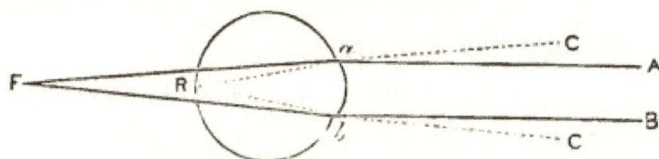

DEF. 5. A *hypermetropic* eye (Fig. 5) is that in which, when at rest, *parallel rays*, (*Aa, Bb,*) come to a focus at *F*, *behind* the retina. Consequently, rays issuing from the retina, *R*, which is within the focus of the dioptric system, emerge from the eye as divergently, (*aC, bC*), as if proceeding from *F*. Further, rays with a convergence as if proceeding to *F*, would unite upon the retina.

NOTE.—The *distance* of this point, *F*, *behind the cornea* will, in any case, be *equal to the focal length of the convex lens, which, when held close to the eye with its accommodation suspended, brings parallel rays to a focus on the retina.* It is evident that the same lens will, conversely, render *parallel* those rays which proceed from the retina. And since a convex lens renders parallel the rays which issue from its principal focus (Chap. ii., *b*) it follows that the rays which emerge from the eye must have a divergence as if proceeding from this point.

DEF. 6. A *regularly* astigmatic eye is that in which there is a difference of refraction in different meridians : the two *principal* meridians being always at *right angles* to each other.

Astigmatism may exist in five different forms.

  a. Simple myopic = one meridian emmetropic the other myopic.

  b. Simple hypermetropic = one meridian emmetropic the other hypermetropic.

  c. Compound myopic = both meridians myopic, one more than the other.

  d. Compound hypermetropic = both meridians hypermetropic one more than the other.

  e. Mixed = one meridian myopic, the other hypermetropic.

DEF. 7. An *Irregularly Astigmatic* eye is one in which there are different degrees of refraction in different parts of the various meridians.

DEF. 8. A *Presbyopic* eye is that in which, owing to physiological changes, produced by advancing age, there is an inability to define small objects nearer to the eye than 22 cm. or about 9 *English* inches, (*vide* Presbyopia chap. xii.).   It may exist with any of the above named conditions.

From what has been stated in this and the preceding chapter we may get the following,

DEDUCTIONS.

Deduction 1. That for *parallel* rays to be brought to a focus on the retina of a myope, they must have given to them a divergence as if proceeding from the "far point" (*vide* def. 4). This is effected by the *concave* lens whose negative focal length equals the distance of this point, (chap. ii. *e.*)

Deduction 2. That to bring *parallel* rays to a focus on the retina of a hypermetropic eye at rest, we require a *convex* lens such that it gives to parallel rays a convergence as if proceeding to the point behind the eye from which rays appear to issue, (p. 9, fig. 5, *F.*)

Deduction 3. That in a regularly astigmatic eye the *two principal* meridians must be corrected by means of cylindrical lenses or a combination of these with sphericals.

CHAPTER IV.

ACCOMMODATION.

DEF. 9. *By accommodation of the eye is meant the power which, by muscular aid, that organ possesses in itself of bringing to a focus on its retina rays proceeding from objects situated at various distances.* These distances range between the "*far point*," for rays from which the eye is adjusted when *not using any* of its accommodation, and the "*near point* for rays from which the *whole* of the accommodative power is called into activity. The amount of accommodation thus exercised by an eye in passing from the former to the latter of these conditions is called the *Range* or *Amplitude* (Donders) of accommodation. This amplitude may be represented by the convex lens which, with *completely suspended accommodation*, would bring rays from the near point to a focus on the retina.

A lens whose *focal length equals the distance of the near point*, will, when placed close to an eye, render *parallel* the rays entering that organ from its near point.

Since in an *emmetropic* eye at rest, *parallel* rays come to a focus on the retina, (definition 3), such a lens represents in *emmetropia* the amplitude of accommodation.

But in a *hypermetropic* eye at rest *parallel* rays do *not* come to a focus on the retina without the aid of a *convex* lens, (deduction 2, chap. iii). To get the amplitude of accommodation the strength of the lens thus necessary for parallel rays must therefore be *added* to that whose focal length equals the distance of the near point.

And in *Myopia*, since a *concave* lens is necessary for parallel rays (chap. iii deduction i.) it is evident that for the amplitude of accommodation, the strength of the lens requisite to bring parallel rays to a focus on the retina must be *deducted* from that whose focal length equals the distance of the near point.

e.g. Let us suppose three eyes, *a.* emmetropic, *b.* hypermetropic 2 D $(\frac{1}{20})$, *c.* myopic 2 D $(\frac{1}{20})$, each having a near point of 10 cm. (4 inches). The amplitude of accommodation will be represented for *a.* by the lens whose focal length = 10 cm. (4 inches) viz. 10 D $(\frac{1}{4})$; for *b.* by the same lens *added to* that which corrects the hypermetropia, viz. 10 D + 2 D = 12 D $(\frac{1}{4} + \frac{1}{20} = \frac{1}{3})$; and for *c.* by the same lens from which is *deducted* the one that neutralises the myopia, viz. 10 D − 2 D = 8 D $(\frac{1}{4} - \frac{1}{20} = \frac{1}{5})$.

From these examples we see that with the *same near point*, the amplitude is *greater* in hypermetropia and *less* in myopia than in emmetropia, and it will be evident that with the *same amplitude* the near point is further from the eye in hypermetropia and closer to it in myopia, than in emmetropia. As age advances the near point recedes further from the eye so that in *presbyopia* there is *absolutely* less amplitude for any given eye. If the near point recedes until it reaches the far point, the accommodation becomes *nil*.

CHAPTER V.

Perception of a Line.

The distinctness with which a line is visible depends upon the sharp and well-defined perception of its *margins;* if these are indistinct the line appears hazy. A line may be taken as made up of an infinite number of elements or points, from *each* of which rays issue in *all* directions. To gain a distinct image of any line, it is necessary that the rays from these points, which emerge in planes at *right angles* to its long axis, should be brought to a focus at points on corresponding planes of the retina, otherwise circles of diffusion are formed in this *transverse* direction of the line, which overlap each other, and finally, by projecting beyond its *margins*, give to it an ill-defined and blurrred outline, so that no distinct perception of·it is obtained. If, from these same points the rays, emerging in planes *parallel* to the long axis overlap each other, it is only at the two *extremities* of the line that, by projecting, they cause any blurring. The *margins* of the line, thus not being in any way affected, there is no interference with the outline, and a clear image is formed on the retina.

If then, a patient with his accommodation suspended, and who is emmetropic in one meridian, and myopic or hypermetropic in the other, be placed at 6 m. (20 ft.) from radiating lines of equal definition, he will see most *dis-*

tinctly that line which runs at *right angles* to his *emmetropic* meridian. Rays from points in the *transverse* planes of *this* line will, by passing through the emmetropic meridian, come to a focus on his retina giving a distinct, well-defined image of the *margins*, and hence a clear perception of the line. The line *parallel* to the emmetropic meridian will, at the same time, be the most *in*distinct. Rays from *its* transverse planes pass through the myopic or hypermetropic meridian. They thus come to a focus, in the former case, in front of, and in the latter behind, the retina producing circles of diffusion in these planes, and a consequent bluring of the *margins* of the line, so that no distinct image is obtained.

RULE I.—We have then the rule that in *simple* astigmatism, the patient, at 6 m. (20 ft.) sees, *most distinctly*, the line *parallel* to the plane of his *error of refraction*.

It follows also, that a patient with either compound or mixed astigmatism, will not see *any* line distinctly at 6 m. with his accommodation suspended.

CHAPTER VI.

DESCRIPTION OF TEST-TYPES: METHOD OF EMPLOYING THEM.

TEST-TYPES are divided into—*a.* Those for *near* vision, and *b*, those for *distant* vision. In using them we are obviously dependent on the answers given by the patient. For children, illiterate adults, and impostors, they are therefore inferior to the ophthalmoscope as a means of diagnosing and estimating refraction.

a. TEST-TYPES FOR NEAR VISION.

Those generally in use for this purpose are Jaeger's or Snellen's. The latter are so graduated that each should be read *as far off* as the distance for which it is marked. The smallest should be seen as far off as 50 cm. (1½ ft.) and the largest at 4 m. (12 ft.). These types are given into the patient's hand, and we then note the *smallest* he can read and the *nearest* and *farthest* points at which *it* is distinctly visible. The small types are chiefly useful in testing the accommodation, but they also afford an indication of the presence and amount of myopia (*vide* Chap. xii., Myopia).

b. TEST-TYPES FOR DISTANT VISION.

Snellen's types for this purpose are so graduated that each should be distinctly legible *as far off* as the distance for which it is marked. The largest should be seen at 60 metres, and each succeeding line at 36, 24, 18, 12, 9 and 6 metres respectively (200 ft., 100, 70, 50, 40, 30, and 20).

In testing with these types, we so place the patient that rays issuing from them reach the eye as parallel rays. For *practical* purposes this is obtained at a distance of 6 m. (20 ft.) (*vide* Chap. ii., *a.*) from which point the lowest line should be read by the normal eye without accommodation. For testing astigmatism, we have a fan of radiating lines, all of the same magnitude and definition. These should appear all equally distinct to the normal eye at 6 m. In noting the vision ($=$ V.) we employ a fraction whose *numerator* denotes the *distance* at which the patient stands, and *whose denominator* indicates the *lowest* line which he *can* read. The number of the line is designated by the distance in metres or feet, at which it *should* be legible. Thus normal V. $= \frac{6}{6}$ ($\frac{20}{20}$) or 1; but if at 6 m. (20 ft.) the patient read only the line which *should* be read as far off as 12 m. (40 ft.) we say V. $= \frac{6}{12}$ ($\frac{20}{40}$) or $\frac{1}{2}$. Seeing that our object in testing refraction is always to have the accommodation suspended, we never place the patient nearer the types than 6 m. Should it be necessary, however, in defective V. *from other causes*, we may allow him to approach the types till he can read the largest; if this be at a distance of say 2 m. (6 ft.) then V. $= \frac{2}{60}$ ($\frac{6}{200}$).

c

CHAPTER VII.

INDICATIONS AFFORDED BY USE OF TEST-TYPES.

NOTE.—In the following section the patient, is tested *without* glasses, then :—

Indication 1. If a patient read Jaeger 1 ($=$ J. 1) or Snellen 1 $=$ (Sn. 1) with a good range, and read also $\frac{6}{6}$ ($\frac{20}{20}$) perfectly he is probably *emmetropic*. He *cannot* be myopic though he *may* be hypermetropic: for a hypermetropic patient with active accommodation *could* do this.

Indication 2. If a patient *over* 40 *years of age* read only the larger J. or Sn. types (or *perhaps* even the smaller) but *only on condition that he holds them at a considerable distance :* while at the same time he can read $\frac{6}{6}$ ($\frac{20}{20}$) perfectly, he is simply presbyopic.

Indication 3. If a patient must hold the types *close* to his eye but can then read even the *finest* though he cannot see $\frac{6}{60}$ ($\frac{20}{200}$) he is *myopic*.

NOTE.—A patient may read Sn. 1 as far as 50 cm. (No. 1 Sn. to $1\frac{1}{2}$ ft.) or even Sn. 2 as far as 60 cm. (No. 2. Sn. to 2 ft.) together with some of the larger distance types such as $\frac{6}{36}$ ($\frac{20}{100}$). In this case he is *very slightly myopic.* N.B. Since such a patient can read Sn. 1 *up to within a short distance from his eye* he is thus easily distinguished from the following.

Indication 4. If a patient read only the *larger* series of the small types, and the smaller series not at all, or only very imperfectly : while at the same time the distant vision is very defective, we suspect either some form of *astigmatism* or *hypermetiopia without accommodation*.

In a patient *under* 40 *years of age* it is *probably* the *former*, but in one *over* that age it may be *either*, though it is frequently only the *latter* condition which exists.

Indication 5. If a patient *under* 40 or 45 can read only such large types as Sn. for 4 m. (J. 16) while he can read $\frac{3}{8}$ ($\frac{20}{20}$) he has *paralysis of accommodation*. This may be proved by giving him a strong convex lens such as $+$ 3 D. ($+$ 12 in.) when he will read Sn. 1, or J. 1, at the focal distance of the lens. (*Vide* Accommodation, chap. iv.)

Indication 6. If a patient, whose accommodation is suspended, see, quite distinctly, *one only* of the radiating lines at 6 m. he is *emmetropic* only in the meridian at *right angles* to *this* line, and *either* hypermetropic or myopic in the other. (*Vide* chap. v.)

CHAPTER VIII.

INDICATIONS AFFORDED BY MEANS OF MIRROR ALONE. "DIRECT
METHOD."

IN order that the following indications of the refraction
afforded by this method may be realised, the observer, if
he be not emmetropic, must correct any error of his own
refraction by means of spectacles or a lens behind the sight
hole of the mirror. He should then seat himself opposite,
and 3 or 4 feet away from, the patient who is to gaze
steadily at the *darkened* wall in front of him, looking
towards the left side of observer's head when the left eye
is under examination and *vice versa*.   By this means the
position of the optic disc is brought into the axis of vision
of the observer.   The latter must now throw the light from
his mirror into the patient's eye, and, keeping the fundus
illuminated, should move his head in various directions.

Indication 1.   If he then get nothing more than the red
reflex of the fundus or at most a *blurred* image of the disc,
the eye is *emmetropic* or *very slightly myopic*.   (*Vide* explanation, page 22.)

Indication 2.   If he see the image of the disc and its
vessels moving in the *same* direction as his own head, the
patient is *hypermetropic*.   (*Vide* p. 23.)

Indication 3.   If the vessels in *one meridian only* are visible,
and these move in the *same* direction as the observer's head

there is *hypermetropia* in *one meridian only*, viz., in that which is at *right angles to the one in which the vessels are visible*. (*Vide* explanation, p. 25, and compare with Chap. v.). = *Simple hypermetropic astigmatism.*

Indication 4. If the disc and vessels move in the *opposite* direction to that of the observer's head, the eye is *myopic*. (*Vide* p. 24).

Indication 5. If the vessels are seen in *one meridian only* and moving in the *opposite* direction to the observer, there is *myopia* in the *meridian at right angles* to that in which the vessels are visible. (*Vide* p. 25, and chap. v.). = *Simple myopic astigmatism.*

Indication 6. If the vessels are seen moving in one direction in one meridian, and in the opposite direction in the other meridian according to the accommodation of the observer, and his distance from the patient, there is *mixed astigmatism.* This condition is, however, not easy of detection by the mirror alone.

Indication 7. If instead of the vessels of the disc moving evenly and regularly, they move slowly across the centre of the pupil, but rapidly and irregularly towards the periphery, giving the appearance of rotating bent spokes of a wheel, the patient has *irregular astigmatism.*

EXPLANATIONS.

The above mentioned indications may be explained in the following manner.

Of all the rays, (*vide* figs. 6 and 7,) which diverge from any *single point, a* or *b,* of the fundus *one only,*

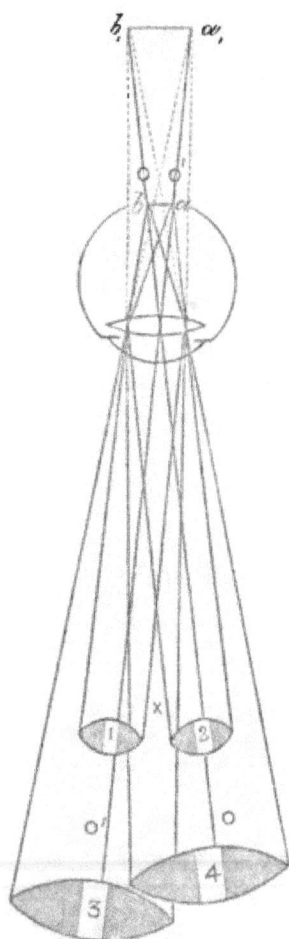

O' or O, passes *without deflection* through the *centre* of the crystalline lens. As many of the re-mainder as the size of the pupil will allow pass out in a cone of rays having a certain relation, according to the refraction, to the ray just mentioned. In order to get a view of any part of the fundus it is necessary that rays from *both* extremities of such portion should come to a focus on the observer's retina.

In *emmetropia*, Fig. 6, the rays which issue from the extremities *a* and *b* of the disc, emerge as two cylinders (1 and 2) of *parallel* rays taking the same direction respectively as those o' and o which from the two extremities *a* and *b* pass through the centre of the lens. There are thus parallel rays in two cylinders, one from either extremity of the disc, emerging from the pupil and soon diverging from each other. They thus leave between them an area x in which there are no rays from these two points. If the observer's eye be situated in this space it is evident that he cannot get an image of the *whole* disc. He will however receive *parallel* rays from some luminous *point* of the disc: or even cylinders from *two very contigu-*

Fig. 6.

ous points may run so closely together that on entering the observer's eye they might *theoretically* form an image of the space between them.   *Practically* however such an image is rarely, if ever, obtained because the *absolute* suspension of accommodation necessary for its production is scarcely to be met with.   We may therefore assume that for practical purposes no details of the fundus are visible with the mirror alone at some distance, and that therefore in *emmetropia* the image of the disc is *blurred*.

In *hypermetropia*, Fig. 6, the rays issuing from the extremities *a. b.* if the disc emerge from the pupil as two cones (3 and 4) of rays which appear *as if* diverging from points a_1 b_1 situated behind the eye on prolongations backwards of the rays o' and o respectively which from *a* and *b* pass through the centre of the lens.   By the union of the rays thus prolonged back we get an *erect* image a_1 b_1 of the disc a b.   (For the *distance* of this image behind the eye, *vide* chap. iii. note to definition 5).

The cones 3 and 4 being formed of diverging rays, separate from each other only at a considerable distance from the eye.   Some of the rays from each extremity of the disc, or at any rate those from two fairly distant portions of the same, will enter the observer's pupil, and, with the exercise of sufficient accommodation to overcome their divergence, will meet on his retina.

An object may be supposed as situated at the points whence *appear* to issue the rays entering the eye.   The *object* therefore which the observer here appears to see is the *erect image* of the disc.   Since the *object* thus apparently seen is *further* from the observer than the pupil, with which it is, though perhaps unconsciously, compared, it seems to move in the *same* direction as the observer's head, (*vide* explanation, p. 25).

In *myopia*, fig. 7, the rays issuing from the extremities, *a. b.* of the disc, emerge as two cones of rays which *converge* to points, *a' b'* on their secondary axis, o' o, respectively, thus forming in front of the eye an *inverted* aerial image of the disc. If the observer be nearer the eye than where these rays unite, he will not get any image of the disc, since he receives only *convergent* rays which cannot come to a focus on his retina. If he, however, be far enough removed he will see the inverted image, *a' b'* of the disc. The rays, having *converged* to form this image, now *cross* and reach his eye with a *divergence* 3 and 4, *as if* proceeding from an object (the aerial image) situated *in front of* the patient's eye.

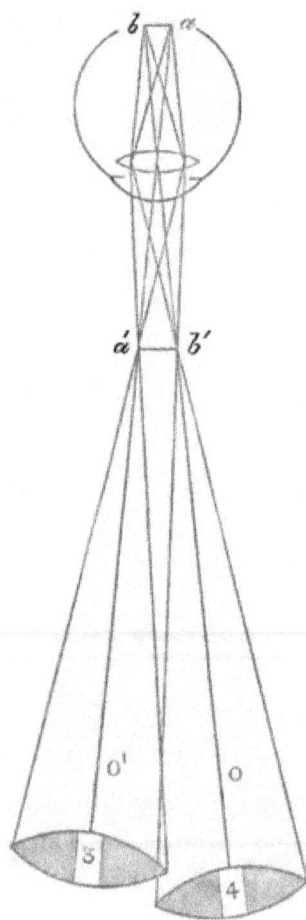

FIG. 7.

The object thus apparently seen, being *nearer* the observer than the patient's pupil, moves in a direction *opposite* to that of the observer's head. (*Vide* explanation, p. 25).

The *apparent* movement of the image of the disc may be illustrated by comparing the patient's eye to an illuminated chamber, outside and opposite to the window of, which the observer stands. The pupil would then be represented by the window, the *image* of the disc in *hypermetropia* by a picture on the opposite wall, and the *aerial image* of the same in *myopia* by some small object situated between himself and the window. If he then moves his head to the right while fixing his eye on the *picture* the latter will *seem* to disappear behind that side of the window which is to his right, *i.e.* it will *appear* to move in the same direction as his head.

But, if while moving to the *right*, he regard now the object which is *in front of* the window, it will finally intervene between him and that side of the window which is to his *left*, and will therefore *appear* to have travelled in a direction *opposite* to that of his own head.

The same explanation as that just given for the whole disc will apply also to vessels seen in different meridians.

In *simple astigmatism*, we do not see vessels at *right angles* to the *emmetropic* meridian because rays from their *transverse* planes (*vide* chap. v.) passing through the emmetropic meridian will (as was seen for simple emmetropia) not be visible. Whereas rays from the transverse planes of vessels at *right angles* to the myopic or hypermetropic meridian will, by passing through these meridians, produce in the former case an inverted, and in the latter an erect, image of the vessels. These follow the same rules of movement as in simple myopia and hypermetropia respectively.

In *compound Myopic astigmatism*, if the observer use his accommodation, vessels at *right angles* to the *most* myopic meridian will be distinctly seen much *closer* to the patient's eye than will those in the opposite meridian. Rays from the transverse planes of the former converge and form an inverted image much sooner than do those from the latter, just as in high degrees of simple myopia.

In *compound Hypermetropic astigmatism* we find that for the vessels at *right angles* to the *most* hypermetropic meridian more accommodation is required at a given distance from the eye, than for those in the opposite meridian. A greater extent of the former than of the latter vessels is, however, at the same time visible. Rays from the former are more divergent and the cones take longer to separate, *vide* Fig. 6.

In *mixed* astigmatism images of the vessels *may* be seen sometimes inverted and at other times erect according to the meridian through which the rays emerge and varying with the observer's accommodation. The details are more visible than· in emmetropia, but, are nevertheless very *indefinite*.

CHAPTER IX.

KERATOSCOPY (RETINOSCOPY).

By means of what has been, unfortunately, called "Keratoscopy," we have a useful method of diagnosing errors of refraction with the ophthalmoscope mirror alone; and, what is, perhaps, equally important, we can correct them with ordinary trial lenses quite independently of aid from the patient. As long ago as 1864 (*vide* "Anomalies of Refraction and Accommodation of the Eye," Donders, p. 490) Mr. Bowman drew attention to "the discovery of regular astigmatism of the cornea, and the direction of the chief meridians, by using the mirror of the ophthalmoscope. . . . . The area of the pupil then exhibits a somewhat linear shadow in some meridians rather than others."

Dr. Cuignet, of Lille, seems to have first systematised this method of examination, and, in 1874, published his conclusions in an article entitled "Keratoscopy."

Dr. Forbes in the *Ophthal. Hosp. Rept.*, Vol. x., p. 62, has lately drawn attention to the subject in this country. It would, I think, be preferable, as suggested by Dr. Parent, (*Recueil d' Ophthal.*, Feb., 1880) to adopt the term "Retinoscopy." The appearances, to be presently described, are due to the play of light and shade on the *retina*. The shadows which we recognise are not formed in the cornea, though this structure, by its different curvatures, as in astigmatism, undoubtedly exercises an influence on their production, and the manner in which we see them.

Rays of light from a distant lamp, falling on a concave mirror, issue from the latter convergingly, and, crossing where they form an image of the lamp in front of the mirror, again diverge.

If, in front of a screen, we place a convex lens at such a distance that diverging rays from a concave mirror are brought to a focus exactly on the screen, there is formed the smallest and brightest possible image of the lamp, and the most sharply defined and densest surrounding shadow. If, then, we shift the lens, it is found that the further it is removed from the point just mentioned, either towards or away from the screen, the *larger* becomes the area of light and the *feebler* is the illumination.

The increasing circles of diffusion render indistinct the line of demarcation between light and shade, and cause the latter to appear fainter. If, with the lens at different distances from the screen, the mirror be variously rotated, the area of light and shade will be seen, in *every* position of the lens, to move, *on the screen*, in a direction *opposite* to that in which the mirror is rotated. If we replace the screen and lens by the retina and dioptric system of the eye (p. 8, DEF. 1) we have precisely similar results. The area of illumination on the *retina*, in *all* states of refraction, really moves in a direction *opposite* to that in which the mirror is rotated. But since this illuminated portion is seen through the transparent media of the eye, the *apparent* direction of its movement will be influenced by the refraction of the eye under examination.

For making use of these facts in practice, atropine though not essential, certainly renders material assistance. The observer should be seated opposite, and 1 m. 20 (48 in.) away from, the patient: the room should be darkened, and the patient's eye shaded by a screen. The mirror used must be *concave*, and should have a focal length of about 22 cm. (9 ins.). If the observer be not emmetropic, he must correct his own error. The patient should regard the opposite wall, while the light from the mirror is thrown into his eye at an angle of 10° or 15° with his axis of vision.

The mirror is to be rotated in different directions, and we then see, in the pupillary area, an *image* of the illuminated and shaded portion of the retina.

If the rays issuing from the observed eye do not cross before reaching the observer, an *erect image* of this light and shade is obtained. This image is seen to move in the direction actually taken by the illuminated area on the retina, viz., *opposite* to that in which the mirror is rotated. This is the case in hypermetropia, emmetropia, and *weak* myopia.

If, however, the issuing rays *cross* before reaching the observer, and form, between him and the patient, an inverted *aerial image* of the illuminated and shaded portion of the fundus, then, since the illuminated area on the *retina* · *really* moves in the *opposite* direction to the mirror, the *aerial image* of the same will appear to move in a contrary sense, *i e.*, in the *same* direction as the mirror. This appearance is met with in cases of myopia of 1 D. ($\frac{1}{40}$) and upwards: for if the observer, with good power of accommodation, be seated as stated, at 1 m. 20, from the patient, then, where the myopia $= 1$ D. ($\frac{1}{40}$) the aerial image being formed at 1 m. (40 ins.) from the patient, and 20 cm. (9 in.) from the observer, will be easily perceived by the latter. Still more is this the case where, with increasing myopia, the image is formed nearer the patient. If the eye under examination be, to such an extent, myopic that its "far point" is situated at the position of the image of the lamp, say 25 cm. in front of the mirror, or 95 cm. from the eye, then the diverging rays from this image will come to a focus exactly on the patient's retina (DEF. 4, p. 9) and there form the smallest and brightest possible image of the lamp.

The further the departure from this degree, which we may call 1 D. ($\frac{1}{40}$) of myopia, the larger and feebler, as was seen on the screen, will be the illuminated portion of the retina. The shadow will, at the same time, become correspondingly fainter, with its line of demarcation less defined.

The difference in degree of luminosity of this reflex can be easily appreciated, so that it furnishes a means whereby the amount of error may be approximately estimated. An attempt has been made to use, for the same purpose, the difference in intensity of the shadow, but there are so many antagonistic factors affecting our perception of its definition and density, that this latter does not afford an altogether reliable source of information.

If, from some distance, say 1 m. 20, we examine an eye with the mirror alone, we find that, the higher the hyper-

KERATOSCOPY (RETINOSCOPY). 31

metropia or myopia, the smaller is the image which we
obtain of the disc: so that, in very high degrees, we see
not only the whole disc, but also some of the surrounding
fundus in the pupillary area. In emmetropia, on the other
hand, so large is this image that only a small portion of it
is visible at one time. With equally rapid rotations of the
mirror then, the light would have to travel much faster
over the large image of the latter, than the small image
of the former, condition. This difference in the *rate* of
movement in the various states of refraction, was pointed
out by my friend Dr. Charnley: it constitutes probably the
best marked and most reliable of the means at our dis-
posal for estimating, by this method, different degrees of
error.

In emmetropia and the lower degrees of hypermetropia
and myopia, so little is seen of the large image of the illuminated area and the surrounding shadow, that the small
portion of the latter, visible at any one time in the pupillary area, appears approximately *linear;* while, although
with the increasing degrees of H. or M., the nearly circu-
lar area of illumination on the retina also enlarges, yet, so
much diminished is the image of the same, that more of
the circumference of the surrounding shadow is visible at
one time in the area of the pupil; it appears, therefore,
more *crescentic*, while it becomes at the same time narrower.

Taking then the *direction* of the movement as denoting
the *kind* of error; and the *rate* of movement, degree of
luminosity, and curvature of shadow as indicating approxi-

mately the *amount* of the same, we may summarise as follows :—

1. If the image of light and shade moves in the *same* direction as that in which the mirror is rotated : and, if the rapidity of movement and curvature of shadow are the same in all meridians, we have a case of *simple myopia*.

2. The same conditions, but with the image moving in a direction *opposite* to that in which the mirror is rotated, indicate either hypermetropia, emmetropia, or *weak* myopia.

3. The *slower* the movements of the image, the *feebler* the illumination, and the more *crescentic*, and narrower, the shadow, the *higher* is the hypermetropia or myopia.

NOTE.—Between 1 D. $\left(\frac{1}{40}\right)$ of myopia and emmetropia, the convergent rays issuing from the eye, unite either behind the observer, or so close in front of him, that no distinct image is received on his retina.   In these cases, the *direction* of the movement, as will be presently explained, taken with the indistinctness, give the best indication of this condition.

4. A *difference*, in two opposite meridians, of rapidity, or direction of, movement, or curvature of the shadow indicates *astigmatism*.   These two dissimilar shadows, moving at right angles to each other, either one vertically and the other horizontally, or both obliquely, denote the meridians of greatest and least refraction.

Since, however, as we have seen, with the shadow moving in the *opposite* direction to the mirror there may be one of three conditions, it becomes necessary to distinguish between them in the following manner. *a.* If, with a convex 1 D ($\frac{1}{40}$) placed by means of a spectacle frame, in front of the observed eye, the shadow *continues* to move in the *opposite* direction, it shows that the patient is *hypermetropic.* But if it now move in the *same* direction, then:—*b.* Place in front of the eye a convex 0·50 D ($\frac{1}{80}$) when, if it were previously myopic 0·50 D ($\frac{1}{80}$), it will now be made myopic 1 D ($\frac{1}{40}$), so that, at the distance of 1 m. 20, we shall get an inverted aerial image of the fundus, as before explained, moving in the *same* direction. *c.* If, with + 1 D ($\frac{1}{40}$) the shade moves in the *same* direction, while with + 0·50 D ($\frac{1}{80}$) it *continues* to move in the *opposite* direction, the observed eye is *emmetropic.*

The *amount* of error may be diagnosed, and corrected, by ordinary trial lenses. Suppose we see the shadow moving in the *same* direction, it proves the presence of *myopia.* If, on placing in front of the patient's eye, a concave 3 D ($\frac{1}{12}$), the shadow still moves in the *same* direction, we try with concave 4 D ($\frac{1}{10}$). If now it moves in the *opposite* direction, we know that the myopia is not *more* than 5 D ($\frac{1}{8}$); for if it were, the concave 4 D ($\frac{1}{10}$) would have left 1 D ($\frac{1}{40}$) uncorrected, so that, at 1 m. 20, the shadow would still have moved in the *same* direction. The myopia is therefore between 4 D and 5 D ($\frac{1}{10}$ and $\frac{1}{8}$).

We proceed in the same manner, with convex glasses, for hypermetropia, until we find the shadow moving in the *same* direction, we then know that there is produced artificial myopia of, at least, 1 D. ($\frac{1}{40}$). The amount of H.

in any case will therefore be about 1 D. ($\frac{1}{40}$) less than the
lens producing the result just mentioned.

In astigmatism, if, by the means just noted, we find one
meridian emmetropic, the other may be corrected by a
cylinder with its axis at *right angles* to the direction taken
by the shadow, since this latter traverses the pupillary area
in a meridian *parallel* to the error which it indicates.

In compound or mixed astigmatism, one meridian is corrected by a spherical lens, and the other by an additional
cylinder with its axis at right angles to the direction taken
by the shadow, which indicates the still uncorrected meridian.

Care must be taken not to throw the light too obliquely
on the eye, for, owing to the obliquity with which the rays
traverse the media, the refraction of the horizontal meridian would be increased.   It must also be carefully noted
that the direction of movement of the *shadow*, as compared
with the *mirror*, is the *reverse* of that of the *disc* or vessels,
as compared with the *observer's head*, (*vide* Chap. VIII.).

CHAPTER X.

INDICATIONS AFFORDED BY MIRROR AND OBJECT LENS.
"INDIRECT METHOD."

THIS method of diagnosing hypermetropia and myopia
was first pointed out by Mr. Hutchinson (*Ophth. Hosp.
Rep.* vol. iv, p. 189) and Mr. Couper (*Med. Times and
Gaz.* Jan. 30, 1869), has shewn how, by it, astigmatism
may also be recognized. It is desirable that the patient
should be under atropine, for otherwise an alteration in the
pupil may easily lead to the supposition that the disc has
changed in size. Upon seeing a change in the shape of
the disc great care is requisite to say whether this is due
to an elongation in one meridian or a diminution in the
other.

In this mode of examination the observer must place
the object lens *as close as possible* to the patient's eye; then,
keeping the optic disc steadily in view, he must gradually
withdraw the lens to the distance of a few inches: when—

Indication 1. If the disc remain the *same size* throughout,
the patient is *emmetropic* (*vide* expl. p 36.)

Indication 2. If the disc *diminish* in size on withdrawing
the lens the patient is *hypermetropic*, and the more so in proportion to the rapidity of diminution (*vide* expl. p. 37).

Indication 3. If the disc *decrease* in size *in one meridian only*,
the patient is *hypermetropic* in *that* meridian only = Simple
hypermetropic astigmatism.

Indication 4. If the disc diminish in *every* direction, but *more* in *one* meridian than the others, there is *hypermetropia all round*, but *most* in *that* meridian which decreases *most* = Compound hypermetropic astigmatism.

Indication 5. *Increase* in size on withdrawing the lens denotes *myopia*, and more in proportion to rapidity of increase. (*Vide* expl. p. 38).

Indication 6. *Increase* in *one* meridian only shows *myopia* in that meridian alone = Simple myopic astigmatism.

Indication 7. *Increase* in *every* direction, but *more* in *one* meridian, indicates *myopia* in *all* directions, but *more* in that meridian which *increases most* = Compound myopic astigmatism.

Indication 8. *Increase* in *one* meridian, and a *diminution* in the *opposite* meridian denotes myopia in the former direction, and hypermetropia in the latter = Mixed astigmatism.

Explanation of change in size of the disc.—These changes in size of the image of the disc will be evident on remembering that *the relative sizes of image and object are as their distances from the lens.* To find the distance of the image from the lens, we have the formula $\frac{1}{b} = \frac{1}{f} - \frac{1}{a}$ where b, = the distance of the image from the lens, f, = the focal length of the lens, and a, = the distance of the object. In each of the following examples the focal length of the lens is 4 cm.

In *emmetropia* the rays issuing from the disc emerge from the eye *parallel* as if proceeding from an object situated at an *infinite* distance (Def. 3). It is of this supposed *object* that we get an *image* by means of a convex lens held in front of the patient's eye. It matters not, therefore, where this lens is held, the *parallel* rays from the object will *always* unite to form an image on the opposite side of the lens at its *principal focus* (Chap. ii., *b*). The relative distances of image and object from the lens remaining *constant*, the size of the image of the disc in emmetropia does not vary with movements of the lens.

In *hypermetropia*, fig. 8,° rays issuing from the disc emerge from the eye as if proceeding from an object, *b*, *a*, at a certain distance behind it (Chap. iii., Def. 5).

It is of this supposed object *b*, *a*, that we gain an *image* I or I' by means of a lens held in front of the patient's eye.

Suppose this lens, *L*, to be at 6 cm. from the *object*, we find by the formula that the distance of the image I is 12 cm. $(\frac{1}{b} = \frac{1}{4} - \frac{1}{6} = \frac{1}{12}$ or $b = 12)$.

If we now withdraw the lens from the object (and also from the eye) till it is 12 cm. from the former as at L' fig. 8, the distance of the image, I' will then be 6 cm. $(\frac{1}{b} = \frac{1}{4} - \frac{1}{12} = \frac{1}{6}$ or $b = 6)$.

The ratio of the distance of the image from the lens as compared with that of the

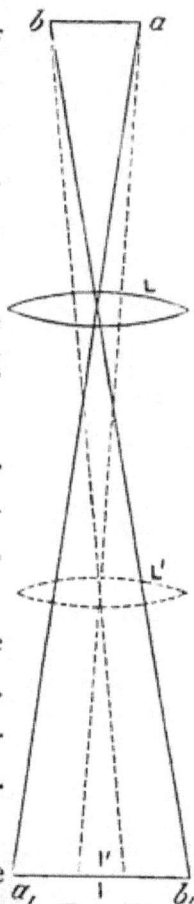

FIG. 8.

* In Figs. 8 and 9, the distances of the lenses from the objects and images are drawn to scale ½ cm. = 1 cm.

object from the lens being *greater* in the first case than in
the second, so is the size of the image. In
hypermetropia on *withdrawing* the lens from the
eye, the image of the disc *diminishes* in size.

In *myopia*, fig. 9, rays from the disc emerge
from the eye convergently and form, at a
certain distance in front of it, an inverted
aerial image, *a*, *b*, of the disc. This latter
we have now to regard as the *object*, o,
whose image, *I* or *I'*, we obtain by means of
a convex lens, *L*, or *L'*, interposed between it
and the patient's eye, *E*. A lens, thus placed,
intercepts the rays, *Bb*, *Aa*, (which, to avoid
confusion, are the only ones shown of all
those) which converge towards the points
b and *a* respectively of this object, o. The
rays being thus rendered more convergent,
produce an image *I* or *I'* whose extremities will be bounded by the lines, passing

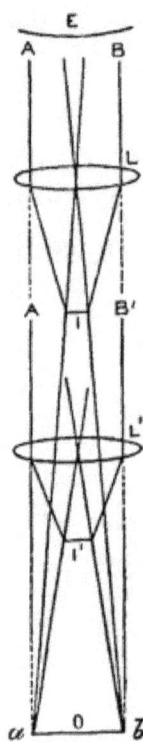

FIG. 9.

from either end respectively of the object, o, through the
centre of the lens, *L* or *L'*.

In this case, object and image are on the *same* side of the
lens. The object here, being on the opposite side of the
lens from the direction in which the rays proceed, it is
customary, in order that the same formula may hold good
in all cases, to regard $\frac{1}{a}$ as a *negative* quantity. For the
distance of the image from the lens then, we have the
formula $\frac{1}{b} = \frac{1}{f} - \left(-\frac{1}{a}\right) = \frac{1}{f} + \frac{1}{a}$.

If we first place a lens, L, 12 cm. nearer the eye than where the object o would be formed, we have the image I at 3 cm. from the lens $\left(\frac{1}{b} = \frac{1}{4} + \frac{1}{12} = \frac{1}{3} \text{ or } b = 3\right)$. If we now withdraw the lens *from* the eye, *i.e., towards* the *object* till it is within 6 cm. of the latter, as at L', the distance of the image I' will be nearly 2 cm. $\left(\frac{1}{b} = \frac{1}{4} + \frac{1}{6} = \frac{5}{12} \text{ or } b = \right.$ nearly 2). But as the ratio of the distance of the image from the lens, as compared with that of the object from the lens, is greater in the second than the first of these examples, so is the size of the image. In myopia, therefore, on approaching the lens to the eye the image diminishes in size, and on *withdrawing* it *from* the eye the image *increases*.

The above explanation holds good for *myopia* only so long as the lens is not withdrawn beyond the " far point" of the eye *plus* its own focal distance: for *hypermetropia* only so long as the focal power of the lens is greater than the degree of hypermetropia. The exception to this latter condition does not occur in practice if a lens of 3 inch focus be used and need not therefore be considered. In the case of very high degrees of myopia, however, if the lens be further from the eye than the aerial inverted image of the disc *plus* its own focal distance, an erect image of the disc is formed between the lens and the observer, the variations in size of which are subject to the same rules as those described for the inverted image in hypermetropia.

CHAPTER XI.

ESTIMATION OF THE REFRACTION BY MEANS OF LENSES IN THE OPHTHALMOSCOPE.

IN estimating refraction by this method the patient and observer must be seated side by side facing in opposite directions. Their heads and the lamp should be on the same level; the latter being placed on the same side of the patient as that under examination, a little out from, and slightly behind the position of, the patient's ear. If now the heads be inclined laterally towards each other, the eyes of the corresponding sides will come opposite one another while the noses and mouths will be left free for breathing. By this means the sight hole of the mirror may generally be placed in the position which, when ordered, the glasses will occupy; the distance of these from the eye need therefore not be taken into consideration when prescribing from measurements thus obtained. Should it however be undesirable to approach thus close, the distance from the patient must be taken into account and the glasses ordered will have to be somewhat *weaker* in *myopia* and *stronger* in *hypermetropia* than the lens necessary to see the fundus would indicate. In estimating refraction by the ophthalmoscope it is absolutely essential that the accommodation of both patient and observer be *completely* suspended. In order to secure the former we must place the patient opposite a *dark* wall or curtain, or employ atropine.

For the observer voluntarily to relax his own accommodation while looking at an object so near to him as the
patient's eye, requires much practice. The best way of
overcoming the difficulty is to regard the fundus as situated some hundreds of yards distant. This habit may be
acquired by looking alternately at a spot on the window
pane and then at some remote object: or by gazing
vacantly at the page of a book until the type disappears;
then, having noticed the sensation produced on thus relaxing the accommodation, endeavour to induce the same
condition while observing a fundus.

In the following indications it is of course necessary that
the observer be fully cognisant of the state of his own refraction, and in the following cases he is supposed to be
emmetropic, (if not *vide* note, p. 43). With accommodation then entirely suspended, the eye of the emmetropic
observer is adjusted for *parallel* rays.

Indication 1. If therefore he can see distinctly the details
of any fundus, the rays issuing thence must be *parallel* and
the eye *emmetropic*.

NOTE. Though this *appears* to contradict what has been
said concerning seeing the disc in emmetropia, (chap. viii,
p. 22), yet we must remember that when *close* up to the
patient's pupil, the observer *does* receive rays from the two
extremities of the disc. Moreover if the observer endeavour to accommodate for the patient's eye, he will find it
impossible to do so, since, if he is as close as he should be,
the latter, as explained by Mr. Power, is nearer to him
than his "near point". The effort is therefore not main-
tained and as when the accommodation is *partially* sus-

pended the disc *begins* to appear, it is unconsciously *altogether* relaxed and a *distinct* view is obtained.

Indication 2. If however the details of the fundus cannot be seen except by the intervention of a *concave* lens, this proves the patient to be myopic. The converging rays from such an eye, (definition 4), require a *concave* lens to render them *parallel* for perception by the emmetrope. The strength of the lens which effects this change of direction indicates the *degree* of myopia, for it is evident that the same lens which renders parallel, the converging rays issuing from the retina, will bring parallel rays to a focus on the retina of the same eye.

Indication 3. If again, with suspended accommodation, the details of the fundus are not visible without the aid of a *convex* lens, this indicates the patient to have hypermetropia. The diverging rays from a hypermetropic eye require a convex lens to render them parallel. The strength of the lens necessary for this purpose denotes the amount of hypermetropia: for the lens which renders parallel, rays proceeding from a retina, will also bring parallel rays to a focus on the same retina.

Indication 5. If the vessels in one meridian are seen without any lens while a convex or concave lens is required for those in the other meridian, it is a case of *simple astigmatism*. The emmetropic meridian is the one at *right angles* to the vessels seen without any lens, (*vide* chap. v, p. 14).

Indication 6. If for the two opposite meridians we require *either* two convex *or* two concave lenses of *unequal* power we have to deal with a case of *compound* astigmatism: hypermetropic or myopic respectively. The *greatest* error of refraction in each case is at *right angles* to the vessels for which the *strongest* lens is necessary.

Indication 7. If for vessels in one meridian we require a *convex* lens, while for those in the other a *concave* is necessary, it indicates *mixed* astigmatism. The *hypermetropic* meridian is at *right angles* to the vessels seen best with a *convex* lens : and the myopic to that for which a concave lens is necessary.

In all these forms of astigmatism then, we see that the lens which, with *absolutely* suspended accommodation, gives the best view of vessels in *one* meridian, measures the *amount of* error which exists in the *opposite* meridian.

NOTE. Should the observer not be emmetropic, yet, if he correct his own error with spectacles, he is in the same position as an emmetropic observer. If he however prefer to estimate refraction without his glasses he must make an allowance for his defect by means of the lenses behind the sight hole of the mirror. To see the details of an *emmetropic* fundus therefore a hypermetropic or myopic observer *starts* with a lens corresponding to the amount of his own error. For any particular fundus then the glass required by him must be to *that* extent *more* convex or concave respectively than would be necessary for an emmetrope to see the same fundus. Consequently the *patient* will always have that amount *less* hypermetropia or myopia respectively than the same lens would indicate if the observer were emmetropic ; we have then the

RULE II. That the observer must *deduct* the amount of his own hypermetropia or myopia from the lens which enables him to see distinctly any particular fundus.

Tabular view of the State of Refraction as indicated by the various methods.

	TEST TYPES.			OPHTHALMOSCOPE.				REFRACTION.
DIST. V.	NEAR VISION.	RAYS IN FAN.	MIRROR ALONE. Movements of Image of O.D. or Vessels.	Movements of Image of Shade.	INDIRECT METHOD. Image of O.D. on withdrawing lens.	Lens necessary in Ophthalmoscope without accom.		REFRACTION.
$\frac{6}{6}$ $\left(\frac{20}{20}\right)$	Sn. 1. 5 in. to 20 in.	All clear.	Red reflex only.	Opposite.	Constant in Size.	None.		E.
$\frac{6}{6}$ by accom.	Sn. 1. or larger, but not close up.	All clear by accom,	Erect, moves in same direction.	Opposite.	Decreases.	Convex.		H.
Not $\frac{6}{6}$	Sn. 1. (or 2) but not so far off as dist. marked.	None.	Inverted, moves in opposite direction.	Same (except for low myopia).	Increases.	Concave.		M.
Not $\frac{6}{6}$	Larger series, (or smaller with difficulty.)	One clear at rt. ang. to Em. merid.	Vess. parl. to Em. merid. in same direction.	Both opposite, (diff. rates.)	Decreases in one merid. only.	Vess. Parl. to Em. merid. = + Vess. Parl. to Hc. merid. = O.		H. As.
			Vess. parl. to Em. merid. in opposite direction.	One opposite. One in same.	Increases in one merid. only.	Vess. Parl. to Em. merid. = — Vess. Parl. to Mc. merid. = O.		M. As.
Variable.	Larger only.	Varying with accom.	Vess. at diff. dists. all in same direction.	Both opposite, (diff. rates.)	Decreases all round, but more so in one meridian.	Two diff. convex.		C.H. As.
		None.	Vess. at diff. dists. all in opp. direction.	Both same, (diff. rates.)	Increases all round, but more so in one meridian.	Two diff. concave.		C.M.As
		One at rt. ang. to Hc. merid. by accom.	Some vess. in same and others in opp. direction.	One opposite. One in same.	Decreases in one but Increases in opp. meridian.	One merid. + Opp. merid. —		Md. As.

Abbreviations: —E. Emmetropia; M. Myopia; H. Hypermetropia;
H. As. Simple Hypermetropic Astigmatism; C.H. As. Compound Hypermetropic Astigmatism;
M. As. „ Myopic „ C.M. As. „ Myopic „
Md. As. Mixed Astigmatism; Vess. Vessels; Parl. Parallel; Diff. Different; O. D. Optic Disc,

CHAPTER XII.

Test Glasses and Types as applied to the Estimation of Refraction.

When ascertaining with the distance types the state of the patient's refraction, we must be absolutely certain that his accommodation is completely relaxed. In myopes, it generally is so, but in hypermetropic and astigmatic patients it is frequently a troublesome and misleading factor. If from the varying and inconsistent statements of the patient, it is suspected that he does not relax his accommodation, he may do so more easily if he close his eyes between each change of glasses, not opening them until the fresh ones are *in situ*. Another method of assisting the patient is to put him on a pair of + 4 D (10 in.) for ten or fifteen minutes, and then, without removing, gradually neutralise them by stronger and stronger concave lenses placed in front, until those are found with which the patient can see in the distance as well as without any glass. The *difference* then between the concave lens thus required and + 4 D ($\frac{1}{10}$), gives the manifest hypermetropia.

One caution must be especially borne in mind, viz.: *never*, in any doubtful case, to commence testing vision with *concave* glasses, for to see with such lenses it is necessary for

all, except those whose myopia is equal to or less than the
glasses being used, to employ their accommodation, and
when once called into activity, this is not easily suspended.

Let the patient be placed at 6 cm. (20 ft.) from the
types. We have already seen what, with test types, are
the indications of emmetropia and simple presbyopia,
(p. 18).

HYPERMETROPIA.

If the patient read $\frac{6}{8}$ ($\frac{20}{20}$) perfectly without glasses, still
this fact does not exclude hypermetropia, the presence of which
is *proved* if he can read the distance types as well with, as
without, *convex* lenses. The *strongest* which are thus toler-
ated indicate the degree of *manifest* hypermetropia, and
they *at least* must be ordered for close work. In young
adults, and in the higher degrees of hypermetropia, we
must for this purpose even give glasses 1 D or 1·50 D
($\frac{1}{40}$ or $\frac{1}{24}$) stronger. In order to rest the ciliary muscle,
it is advisable for distant V. to prescribe at any rate the
glasses which correct the *manifest* hypermetropia. If the
H. be measured under atropine, the glasses ordered must
be 1 D. (or perhaps even 2 D.) *weaker* than the *total* amount
thus found. These should be suitable for all purposes,
though, for distant V., weaker ones are sometimes necessary.

MYOPIA.

As we have already stated, (Chap. VII. Ind. 3) if the
patient *can* read the *finest* type to within 4 in. or 5 in. of
his eye, while at the same time his distant V. does not

exceed $\frac{6}{24}$ ($\frac{20}{70}$) and is probably not $\frac{6}{60}$ ($\frac{20}{200}$), he is *myopic*. An indication of the *degree* may be obtained by observing the *furthest* point at which either of the smallest types can be read. Provided it be nearer than that for which it is marked, the *distance* of this point gives the *focal length* of the lens required to neutralise the myopia.

e.g.—If Sn. 1 be legible only as far off as 12 cm. (5 in.) the myopia is measured by the concave lens having this focal length, viz., 8 D $\left(\frac{1\,D}{12\,cm.}\right)$ or ($\frac{1}{5}$ in.). Or again, if No. 4 Sn. marked for 1 D, (3 ft.), can be read only at 50 cm., (20 in.) the myopia $= 2$ D $\left(\frac{1\,D}{50\,cm.}\right)$ ($\frac{1}{20}$).

In testing the distant V. we commence with the *weaker* concave glasses, and work up to the stronger. In doing this, we cannot be too careful in ascertaining which is the *weakest* glass that neutralises the myopia and gives the best V., whether this be $\frac{6}{6}$ ($\frac{20}{20}$) or less. The glass thus found gives the measure of the myopia, and may in *all* cases be ordered for *distant* V. Such glasses may also be given for *close* work when, *with good accommodation*, the myopia does not exceed about 6 D or 8 D ($\frac{1}{6}$ or $\frac{1}{5}$). In *most* cases where the myopia is higher, and in *all* where the accommodation is feeble, we must order *weaker* glasses for *close* work. These, according to Donders may be found in the following manner, viz.: From the neutralising lens *deduct* the strength of the glass whose focal length equals the distance at which we wish the patient to work.

e.g.—With myopia $= 10$ D ($\frac{1}{4}$) we wish the patient to do work at 40 cm. (16 in.) From 10 D ($\frac{1}{4}$) we deduct therefore the lens whose focal length is 40 cm. (16 in.),

viz.: 2·50 D ($\frac{1}{16}$), and the glasses ordered will be (10 D —
2·50 D =) — 7·50 D $\left(\frac{1}{4} - \frac{1}{16} = \frac{1}{5\frac{1}{3}}\right)$

For the various exceptions to these, only very general
indications of glasses to be ordered, as well as for the
numerous complications of myopia, the reader must be
referred to the larger works on this subject.

Astigmatism.

If the V. cannot be brought up to $\frac{6}{6}$ ($\frac{20}{20}$) with any *spherical*
glasses, we probably have to do with a case of *astigmatism*.
In dealing with this error *each* eye is to be tested *separately*.

If, from the previous examination with the ophthalmoscope, we have diagnosed,—

(1.) *Simple* astigmatism, we can proceed to test with
the fan of rays. If our surmise has been correct, the
patient should now see *quite distinctly, only* the line at
right angles to his emmetropic meridian, (Chap. V.) If
this line were *vertical*, it would therefore denote that
the meridian *parallel* to it was either hypermetropic or
myopic; we should then ascertain what cylinder, with
its axis at *right angles* to this latter meridian is required
to correct it: *i.e.*, to render clear the horizontal line.
If the correction thus found gives V. = $\frac{6}{6}$ ($\frac{20}{20}$), it may
be ordered for constant use; but if it do not, then,
as where none of the rays are distinct, and more particu-
larly if first one then another is most clearly seen, we must
not hesitate to employ atropine.

It matters not now, whether we have to deal with—

(2.) Compound or mixed astigmatism. We have merely to ascertain what *spherical* lens clears *one* of the rays, and then, leaving this glass *in situ*, try what cylinder, with its axis at *right angles* to the line thus cleared, renders equally distinct the ray in the opposite meridian. This spherico-cylindrical correction, *after due allowance for atropine* (Chap. XIII.), is ordered for constant use.

(3.) Another less scientific, though practically useful plan, is to substitute the test types for the lines, and having found the spherical lens which gives the best V., try what additional cylinder is required.

(4.) If *each* meridian has been measured *separately* with spherical glasses, either for the fan, or in the ophthalmo-scope for the fundus, we shall have to calculate what spherico-cylinder is required. *(a.)* For *compound* astigma-tism we generally give the *spherical* lens, which corrects the meridian of least error, and then add the cylinder (concave for myopic, and convex for hypermetropic as-tigmatism), whose strength equals the difference between the two meridians. *(b.)* In *mixed* astigmatism, the *difference* between the *sphericals* also gives the degree of astigmatism and the strength of the cylinder required. If, therefore, we correct the myopic meridian with a *concave spherical* lens, we shall require in addition a *convex* cylinder and *vice versa*.

e.g.—With vertical meridian myopic 2 D ($\frac{1}{20}$) and the horizontal meridian hypermetropic to the same extent, the *difference* between them = 4 D ($\frac{1}{10}$). If, therefore, we give *minus* 2 D sph. ($\frac{1}{20}$), we must add + 4 D cyl. ($\frac{1}{10}$) with its

E

axis *vertical*, (*i.e.*, at *rt. angles* to the hypermetropic meridian, Chap. II., p. 7.)   But if we give *plus* 2 D sph. ($\frac{1}{20}$),
we shall require in addition — 4 D cyl. ($\frac{1}{10}$) axis horizontal.   The action of atropine (Chap. XIII.), must of course
be taken into account.

5. Another method of testing astigmatism is by means
of Tweedy's optometer, of which, with the exception that
the later instruments, are marked in dioptres and inches
instead of only in inches, he has published a description
in the *Lancet*, for Oct. 28, 1876.   The patient, under atropine, [is made artificially myopic by a convex lens, a card
with fine radiating lines is gradually approached to his
eye, until *one* line becomes quite distinct.   The meridian
at *right angles* to this line is then known to be the *least*
refractive.   The *concave* cylinder is found, which, with its
axis at *right angles* to the line first seen, makes the line in
the opposite meridian equally distinct.   This shews that
the latter meridian is now made as little refracting as the
former.   The distance at which the first line is seen indicates the kind and degree of the error for the *least*
refracting meridian.   A *spherical* lens correcting this is
then ordered, and, in combination with it, the *concave*
cylinder of the strength, and in the axis, which were found
necessary to equalise the two meridians.

PRESBYOPIA.

Though the patient does not require glasses for distant
V., he may, as age advances, find it more and more diffi-
cult to *read* without them. This defect is called *presbyopia*,
and is caused chiefly by failure in the power of accommodation, but is also accompanied by a flattening of the
crystalline lens. It is *indicated* by a *recession of the near
point*, and is said by Donders to have commenced when
this is further from the eye than 22 cm. (9 in.), *i.e.*, than
the focal length of a lens whose refracting power is 4·50 D
($\frac{1}{9}$). As we have already seen, (Chap. IV.) such a lens
would, in *emmetropia*, bring rays from the "*near*" point to
a focus on the retina in the absence of *all* accommodation.

The *difference* then between this lens and one whose focal
length equals the distance of any given *receded* near point,
denotes for *emmetropia* the amount of accommodation which
is deficient, and the lens necessary to enable the patient
again to read at 22 cm.

e.g.—With near point receded to 40 cm. (16 in.) we
must deduct from 4·50 D. ($\frac{1}{9}$) the lens, whose focal length
is 40 cm. (16 in.) viz., 2·50 D. ($\frac{1}{16}$) and have, as the necessary lens, + 2 D. ($\frac{1}{20}$).

When *all* accommodation is lost, even an eye which was
originally emmetropic may require a convex lens for parallel
rays (Donders). The strength of this lens must, of course,
then be *added to* 4·50 D. ($\frac{1}{9}$) in order that such a patient
may read at 22 cm. (9 in.)

The following table gives, according to Donders, the glasses necessary for presbyopia in *emmetropia* at different ages.

Age.	Glass.	
	D.	English Inches.
45	1	$\frac{1}{40}$
50	2	$\frac{1}{20}$
55	3	$\frac{1}{12}$
60	4	$\frac{1}{10}$
65	4·50	$\frac{1}{9}$
70	5·50	$\frac{1}{7\frac{1}{2}}$
75	6	$\frac{1}{7}$
80	7	$\frac{1}{6}$

Since these are the glasses which enable the *emmetropic* eye to see at 22 cm. (9 in.), it is evident that, if either myopia or hypermetropia be present, the amount of the former must be *deducted from*, and that of the latter *added to*, the lens here specified for any particular age. Though these are *theoretically* the glasses required, we must test practically each individual case, for the various patients prefer different distances at which to read or work.

Aphakia.

or "*absence of lens*," as after cataract operations, involves loss of accommodation. If we replace the crystalline lens by a glass in front of the eye which focusses parallel rays on the retina, we render the eye *practically emmetropic*.   To enable such a patient to read at any specified distance, it is necessary merely to add to that glass, the lens whose focal length equals the distance in question.

e.g.—If a patient requires + 13 D. ($\frac{1}{3}$) for parallel rays and we wish him to read at a distance of 33 cm. (12 in.), we add together 13 D. ($\frac{1}{3}$) and the lens whose focal length is 33 cm. (12 in.), viz.: 3 D. ($\frac{1}{12}$), thus getting + 16 D. ($\frac{1}{24}$) as the necessary glass.

For cataract patients we order two pairs of spectacles. One for distant V., generally + 10 D. to + 13 D. ($\frac{1}{4}$ to $\frac{1}{3}$), and the other for reading, from + 15 D. to + 20 D. ($\frac{1}{24}$ to $\frac{1}{2}$) according to the previous state of the refraction.

To ascertain whether the glasses given, accord with our prescription, it must be remembered that if, while moving a *convex* lens to and fro in front of our eye, we regard some distant object, this latter appears to move in the *opposite* direction.   A contrary effect is produced with a concave lens.   A concave and a convex lens of equal strength *neutralise* each other, and there is no movement of the object.

If this latter result is obtained with a lens of the *same* strength as that ordered (though of an *opposite sign*), the glasses are correct.   If the lens, thus necessary, is *not* of the same strength, we ascertain the amount of error by noting the number of the glass required for neutralisation.

CHAPTER XIII.

ATROPINE :—ALTERATIONS NECESSARY IN MEASUREMENTS
MADE UNDER ITS INFLUENCE.

ACCORDING to DEF. 3, parallel rays *should* come to a focus
on the retina of an emmetropic eye when the accommodation is completely paralysed by atropine.   Emmetropia,
however, as thus defined, is rarely met with.   An eye may
have V. = $\frac{6}{6}$ ($\frac{20}{20}$) and its distant V. made indistinct by
even the *weakest* convex lens, yet, when fully under atro-
pine, it will generally be found that a certain amount of
accommodation *was* being exercised, for a *convex* lens is
now necessary to see $\frac{6}{6}$ ($\frac{20}{20}$), *i.e.*, to bring practically
parallel rays to a focus.   That portion of the accommoda-
tion which can be overcome only by atropine is due to the
" *tone* " of the ciliary muscle, and must always be taken
into the calculation when ordering glasses from the measurement of the refraction as determined under atropine.
For *practical* purposes, we may consider that eye emmetropic, which, when fully under atropine, does not require a
convex lens stronger than 1 D ($\frac{1}{40}$) for parallel rays.

In *Hypermetropia* the convex lens which brings parallel
rays to a focus, when the eye is fully under atropine, should
theoretically convert it into an emmetropic eye.   It is found,
however, *practically* that, owing to hypertrophy of the ciliary
muscle, patients with this error of refraction find great diffi-

culty in relaxing their accommodation to anything like its fullest extent. The result of this is, that the lens which brings
parallel rays to a focus under atropine, cannot be ordered
for the same eye with its accommodation active. The
increased convexity of the crystalline lens (produced by
accommodation) added to the correcting lens just mentioned, renders parallel rays too convergent, and brings
them to a focus in front of the retina. In hypermetropia
then, it becomes necessary to *deduct* from the measurement
made under atropine at least 1 D. ($\frac{1}{40}$) and in children and
young adults, frequently as much as 2 D. ($\frac{1}{20}$) for the tone
of the ciliary muscle.

In *myopia* the concave lens which brings parallel rays to
a focus on the retina of an eye under atropine will be
found *too weak* for the same eye when *not* under atropine.
The effect of the accommodation is to bring to a focus in
front of the retina, the rays which, under atropine, were
focussed upon it: in other words, the myopia is still uncorrected, and requires a concave lens 0·50 D. ($\frac{1}{80}$) or 0·75
D. ($\frac{1}{50}$) stronger. We have, therefore, the

RULE III.—That glasses ordered for permanent use
must, when *convex*, be 1·50 D. ($\frac{1}{24}$) to 2 D. ($\frac{1}{20}$) *weaker*, and
when *concave*, 0·50 D. ($\frac{1}{80}$) to 0·75 ($\frac{1}{50}$) *stronger* than is indicated by the measurements made under atropine.

The application of this rule need be exemplified only in
a case of mixed astigmatism. If, under atropine, one
meridian be hypermetropic, 2 D. ($\frac{1}{20}$) and the other myopic, 2 D. ($\frac{1}{20}$), the fully correcting glass would be *either*
+ 2 D. sph. $\bigcirc$° — 4 D. cyl. (+ $\frac{1}{20}$ sph. $\bigcirc$ — $\frac{1}{10}$ cyl.) *or* —

* The sign $\bigcirc$ means " combined with."

2 D. sph. $\supset$ + 4 D. cyl. ($- \frac{1}{20}$ sph. $\supset + \frac{1}{10}$ cyl.). The glasses ordered for permanent use would be, in the first case, + 1 D. sph. $\supset - 4$ D. cyl. (+ $\frac{1}{40}$ sph. $\supset - \frac{1}{10}$ cyl.) and in the second — 3 D. sph. $\supset$ + 4 D. cyl. ($- \frac{1}{12}$ sph. $\supset + \frac{1}{10}$ cyl. In the former example, by *deducting* 1 D. ($\frac{1}{40}$) from the *convex spherical* we at the same time *practically add* 1 D. ($\frac{1}{40}$) to the strength of the *concave* cylinder (for it has thus *less* neutralisation to perform). While in the second example, by *adding* 1 D. ($\frac{1}{40}$) to the *concave spherical* we *lessen* the strength of the *convex* cylinder, since it now has more to neutralise.

In *mixed* astigmatism, then, by *deducting* from the strength of the spherical, we *increase* that of the cylinder: and, by adding to the strength of the spherical, we *diminish* that of the cylinder.

In *compound* astigmatism, if we add to, or deduct from the strength of the spherical, we at the same time increase or diminish respectively the correction for each meridian.

The calculations necessary for the action of atropine may thus be made for both meridians simultaneously, by means of the *spherical* lens alone.

In simple astigmatism, as we have seen (p. 48) atropine is not necessary.

CONCLUSION.

IT need scarcely be mentioned in conclusion that there are many cases in which, though the error of refraction has been duly diagnosed, estimated and corrected, yet, there is no consequent improvement in vision. Such results may be found in old standing cases of strabismus or in diseased eyes. Others again, without any error of refraction, may have defective vision owing to disease; with such, however, we have not here to deal. These pages have been intended merely to indicate the various modes of estimating and correcting errors of refraction; the reason why, after such correction, the vision is not improved must be ascertained by other means. Some patients, in whom we cannot find either disease or error of refraction may simulate total or partial blindness. The latter may often be detected by holding in front of their eyes different pairs of convex, neutralised by their corresponding concave lenses. With these, thinking they are being aided, such patients may frequently be persuaded to read perfectly. Another method of detecting simulation, especially of *one* eye, is to hold in front of the *blind* (?) eye, a prism with its base out or in, when, if there be an attempt (seen by movement of the eye) to fuse the *double* images, it proves sight present in the eye in question.

If some such order of procedure, as indicated in these pages, were adopted, one would not, as is now frequently the case, see a beginner fall into the error of supposing a patient myopic because he can read $\frac{6}{6}$ ($\frac{20}{20}$) as well with, as without, a *concave* lens; or that hypermetropia is present because the patient can read Sn. 1 with *convex* glasses.

Some of the statements may have appeared somewhat sweeping but it has been thought advisable not to confuse the beginner by enumerating all the possible and minor exceptions to the general rules.